Making and Selling Cosmetics Sweet Orange Lip Balm
ISBN: 978-1-912271-87-0
Published by Northern Bee Books © 2021
Northern Bee Books, Scout Bottom Farm
Mytholmroyd, Hebden Bridge, HX7 5JS (UK)
www.northernbeebooks.co.uk
Tel: 01422 882751
Front cover image by Jasmine Patel. All other photographs by Dr Sara Robb
To contact Dr Sara Robb: CPSR@drsararobb.info
Design by SiPat.co.uk

Making and Selling Cosmetics

Sweet Orange Lip Balm

Dr Sara Robb

Contents

Foreword 5
Preface 6
Acknowledgements 7
Introduction 7
The Chemistry of Mixtures 8
Cosmetic Balms 9
Ingredients Used to Make Lip Balm 9
Functions of Ingredients in Make Sweet Orange Lip Balm 11
Equipment 11
Sweet Orange Lip Balm Recipe 13
Making the Sweet Orange Lip Balm 13
Selling Cosmetics, The Legal Requirements 16
Responsible Person 17
Product Information File 17
Ingredient Information in the PIF 18
Cosmetic Product Safety Report 19
Submit a Cosmetic Product Portal 19
Labelling Sweet Orange Lip Balm 19
Selling Your Lip Balm 21
References 23
Index 25

Foreword

This edition of Making and Selling Cosmetics is the first to follow the United Kingdom's departure from the European Union. For many small producers, navigating the laws for selling cosmetics after Brexit will be a source of stress. The detailed advice provided here will help guide you through the process and ensure that you comply with the legislation.

Sara

Dr Sara Robb

Preface

Lip balm is the ideal product to start with when you are learning to make cosmetics. A simple mixture made with just a few ingredients ensures you will have a usable balm after your first attempt.

When I introduce beekeepers to making cosmetics, I often begin with lip balm. An almost fool-proof cosmetic, lip balm making builds confidence in the students and prepares them for creating more-complex cosmetics.

The humble balm, although so easy to make, is an incredibly useful cosmetic. Lip balm helps maintain healthy skin, especially in the winter months. Usually, a blend of wax and oil, a simple formulation, might be made with beeswax and almond oil. Beeswax provides a superior barrier to lock in moisture, while almond oil acts as an emollient to soften the skin.

Sweet Orange Lip Balm is a more sophisticated version of the balm above, incorporating beeswax, cocoa butter, rapeseed oil and sweet orange essential oil. The result, a useful, yet yummy cosmetic that is a treat to use.

Dr Sara Robb

Acknowledgements

My interest in lip balms began with my daughter Jasmine's love for the product. When Jasmine was a little girl, she spent countless hours mixing potions with me in the kitchen. We began making simple balm formulations, and through our experimental mixing, we developed a series of lip balm recipes published in Dr Sara's Honey Potions. Little Jasmine was the inspiration for the fairy on the label of Bath Potions' Sugar Plum Kiss, and all these years later, she still enjoys making and using lip balm.

I would like to thank a more grown-up Jasmine for taking the Sweet Orange Lip Balm photo that appears on the cover.

Introduction

The Making and Selling Cosmetics series will take you through the process of producing cosmetics and the steps required to place them on the market legally. The first edition contained a recipe for Honeycomb Cleansing Cream and explained European Union legislation. Since the first booklet's publication, the United Kingdom has left the EU, and some regulations have changed. This booklet outlines those changes.

My recipes are written with the small producer in mind- formulations that can be successfully made at home using kitchen equipment and easily obtainable ingredients. What's more, these kitchen cosmetics will pass safety assessment and can be legally sold in the UK.

This booklet concentrates on an easy-to-make cosmetic mixture, Sweet Orange Lip Balm. The formulation is a modification of the Luscious Lip Balm recipe, included in the 2019 BBKA Spring Convention workshop, "Make your own Pampering Potions". Simply follow the steps described below to make and sell your own Sweet Orange Lip Balm.

The Chemistry of Mixtures

One of the skills required to make cosmetics is knowing how to blend ingredients to make a stable, desirable product. The easiest method of manufacture is to stir ingredients together. In chemistry, a product made up of two or more different, physically combined substances is called a mixture. This physical combination maintains the starting substances- no chemical reaction occurs. Uniting the individual ingredients results only in the redistribution of the components, as shown in Figure 1. Some examples of mixtures are a tossed salad, a mixed bag of coloured candy and lip balm.

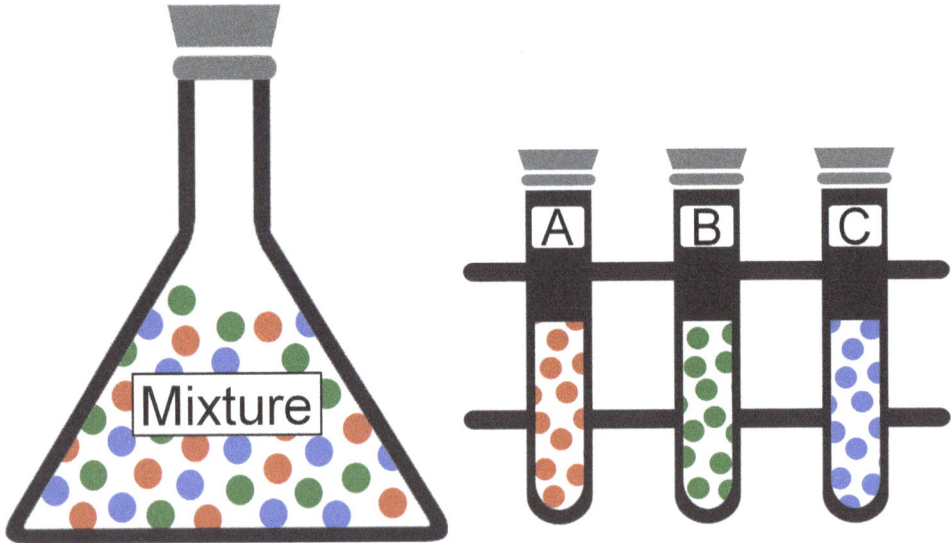

Figure 1. Mixtures result when two or more unique components are blended. The three test tubes each contain a different substance, represented by red, green or blue spheres. The three ingredients are poured into a flask and stirred to redistribute the individual components. The resulting mixture contains all three constituents; red, green and blue spheres distributed evenly throughout the container. This same process occurs when you make a cosmetic mixture like lip balm.

Cosmetic Balms

Anhydrous preparations of wax and oil are called balms. Topically applied, they offer comfort and protection. Manufacturers can customise these cosmetics by selecting different starting materials, creating specialist products, including foot balm, body butter and lip balm. Specific formulations might include colour, flavour or functional ingredients.

Ingredients Used to Make Lip Balm

Lip balms are an example of cosmetic mixtures formulated for a specific use. Protecting and moisturising skin, lip balm contains wax and oil. By incorporating other cosmetic ingredients, you can make varieties. Lip balms frequently contain flavour and colour, making them aesthetically appealing. Like other balms, lip balms contain lipophilic (oil-loving ingredients), including oils, butters and waxes. (Displayed in table 1).

Table 1. *A Selection of Lip Balm Ingredients*

Oils	Butters	Waxes
Rapeseed oil	Cocoa Butter	Beeswax
Sunflower Oil	Shea Butter	Carnauba Wax
Coconut Oil	Mango Butter	Soya Wax
Almond Oil	Petroleum Jelly*	Paraffin Wax
Mineral Oil		Jojoba oil

Lip balms include the components above. To formulate a lip balm recipe, choose one ingredient from each column. Start with equal quantities of oil, butter and wax and adjust until you are happy with the product. *Petroleum jelly is a mixture of mineral oil and paraffin wax and can be used in place of butter in lip balm formulations

Oils, butters and waxes are chemically similar substances from animal, vegetable or mineral origin. Composed of molecules that do not mix with water, they effortlessly combine with other lipophilic substances to create cosmetic mixtures.

We assign the labels of oil, butter and wax to lipophilic substances based on their physical characteristics. Oils tend to be liquid at room temperature, while butters and waxes are often solid. An exception to this is jojoba oil, which is actually a liquid wax. Butters frequently have lower melting temperatures than waxes. Along with having higher melting temperatures, waxes also tend to be malleable. There is some overlap between the categories, yet each ingredient has unique qualities.

A good lip preparation should be firm when touched. However, melt enough from body heat that a thin film can be topically applied. Lip balm formulations are solid at room temperature yet melt at body temperature (37°C) this is achieved by selecting ingredients with different melting points. Sweet Orange Lip Balm combines rapeseed oil, cocoa butter and beeswax. To this base, sweet orange essential oil is added to flavour the finished product. All of these ingredients are in Figure 2.

Formulator's Tip
You may need to adjust your recipe with the seasons. If your lip balm is too firm in the colder months, add a little oil, too soft in the warmer months, add a bit more wax.

Figure 2. *The ingredients used to make Sweet Orange Lip Balm. From left to right, sweet orange essential oil, rapeseed oil, beeswax and cocoa butter.*

Functions of Ingredients in Sweet Orange Lip Balm

In addition to a solid and easily applied cosmetic, there are several other characteristics desirable in a lip balm. Each ingredient included in the formulation conveys different qualities to the finished product. Table 2 summarises the cosmetic functions of the ingredients used in the recipe described here.

Table 2. Functions of Ingredients

Cosmetic Ingredient	Function
Beeswax	Emollient, film-forming, aroma
Cocoa Butter	Emollient, skin conditioning, skin protecting
Rapeseed Oil	Skin conditioning
Sweet Orange Essential Oil	Aroma

Each component of the recipe has a specific function, giving Sweet Orange Lip Balm the qualities listed above.

The characteristics of lip balms depend on the ingredients used in the recipe. Emollients, which smooth and soften, in the Sweet Orange Lip Balm recipe are beeswax and cocoa butter. As well as adding aroma, beeswax is film-forming, leaving a barrier on the surface epidermis. Cocoa butter helps to protect the skin from the environment, such as wind and pollution. Skin conditioning ingredients, including cocoa butter and rapeseed oil, maintain the physical integrity of the epidermis. Finally, the addition of sweet orange essential oil to the base gives the finished lip balm a delicate orange aroma.

Equipment

Beekeepers are likely to have the equipment needed to make cosmetics in their kitchen. If you have a kitchen scale, microwave or hob, heat-proof containers, bowls, and spoons, you can make lip balm. A mini-digital scale that weighs in increments of 0.1 grams (g) is ideal for measuring small quantities as some scales are not accurate below 10 g.

***Figure 3**. Equipment used to make lip balm, including a microwave, weighing bowl, small scales, a Pyrex jug, spoon and whisk.*

***Figure 4**. A selection of some of the packaging available for lip balm, including plastic twist-up tubes, aluminium pots and small plastic jars.*

You will need packaging for your cosmetic products. Choices include plastic pots or tubes, glass jars or aluminium containers. Use the volume of your recipe to calculate how many containers you need. The Sweet Orange Lip Balm recipe makes approximately 150 millilitres (ml) of lip balm or enough to fill ten to twelve 15 ml containers (see Formulator's Tip).

Sweet Orange Lip Balm Recipe

30 grams Beeswax
30 grams Cocoa Butter
90 grams Rapeseed Oil
10 drops Sweet Orange Essential Oil

Formulator's Tip
Use the total weight of your ingredients to estimate the volume of your recipe. The total weight of the Sweet Orange Lip Balm recipe is 150.5 grams (g) which will give you between 150.5 and 180 millilitres (ml) volume.

Making the Sweet Orange Lip Balm

Gather your Equipment

- A scale that weighs in increments of 1 gram or less (ideally 0.1 grams)
- Weighing bowl
- Pyrex jug (if you will use a microwave) or small pan (if you will use the hob)
- Whisk
- Containers (between ten and twelve 15 ml jars- See Formulator's Tip)

Weigh the Ingredients

- Place the weighing bowl on the scale and tare
- Weigh beeswax into the bowl
- Transfer the beeswax to the Pyrex jug or small pan
- Repeat the process with the cocoa butter and rapeseed oil
- DO NOT ADD the essential oil at this stage

Figure 5. *An accurate scale capable of weighing small quantities is ideal for making lip balm. The unit above measures in increments of 0.1 grams and can weigh below 10 grams with precision.*

Figure 6. *The lip balm ingredients; rapeseed oil, beeswax and cocoa butter, before melting.*

Making the mixture

- The beeswax, cocoa butter and rapeseed oil should now be in your jug or pot
- Heat the contents until the solids have melted (take care not to overheat)
- Stir to blend the ingredients
- Add ten drops of sweet orange essential oil and mix thoroughly

Figure 8. Add sweet orange essential oil to the melted ingredients and mixed lip balm base. Stir to incorporate the orange flavour.

Pouring the Lip Balm

- Carefully fill each container
- Take care to fill your containers uniformly
- Leave the lid off until the lip balm has cooled
- When cool, secure the lid and label appropriately

Figure 9. Pour the warm lip balm liquid into the aluminium containers to the fill line. Leave the liquid balm to cool with the lip off.

Selling Cosmetics, The Legal Requirements

Until recently, the requirements to legally sell cosmetics, including lip balm, in the United Kingdom (UK) were the same as those of the European Union (EU). Since Brexit, there are a few changes, but much remains the same.

A new condition resulting from Brexit is that there must be a UK-based Responsible Person (RP) for the product. The RP must be physically present in the UK and is required to put together a Product Information File (PIF) for each cosmetic.

A key component in the PIF is the Cosmetic Product Safety Report (CPSR) issued by a qualified safety assessor. With the RP and PIF in place, the cosmetic product is then listed on the new UK Submit a Cosmetic Product portal. A description, detailed formulation and pictures of the product with labels are listed on the portal. Once submitted, your product is assigned a number, and you can legally sell your cosmetic in the UK.

Responsible Person

The Responsible Person ensures the cosmetic product complies with UK Cosmetics Regulations. For small producers, the manufacturer is usually the RP; however you can also use your business. Additionally, the Responsible Person prepares a Product Information File for each product (every product has its own PIF).

Product Information File

Information contained in the PIF should include; a 1. a description of the product 2. A Cosmetic Product Safety Report (CPSR) prepared by a suitably qualified assessor, 3. statement on animal testing 4. description of the manufacturing methods used, 5. statement of Good Manufacturing Practices (GMP) 6. proof for any claims made, 7. chemical dossiers for all ingredients used to make the cosmetic.

The PIF is held and maintained by the Responsible Person. The RP also prepares a statement about good manufacturing practice, proof of claims and a declaration that no animal testing has occurred. Add an outline of the manufacturing process and detailed information about each ingredient in the formulation.

Ingredient Information in the PIF

> **Formulator's Tip**
> Percentage of an ingredient = grams of ingredient in recipe ÷ total recipe grams x 100
> Beeswax % = (30% ÷150.5g) x 100g = 19.93%

The cosmetic product's ingredients should be specified in the PIF, using the International Nomenclature of Cosmetic Ingredients (INCI), with the percentages of each ingredient used in the formulation. See Table 3 for the ingredient INCI names and amount used in Sweet Orange Lip Balm recipe. Ingredient documentation and results of any testing should follow. Insert International Fragrance Association (IFRA) certificates and allergen declarations. Results of stability testing and results from the microbial analysis are incorporated if they have been undertaken. Microbial testing is usually not required for non-aqueous cosmetics such as lip balm.

Table 3

Ingredient	INCI	Quantity	Percentage
Beeswax	Cera Alba	30 grams	19.93%
Cocoa Butter	Theobroma Cacao	30 grams	19.93%
Rapeseed Oil	Brassica Campestris Seed Oil	90 grams	59.81%
Sweet Orange Essential Oil	Citrus Aurantium Dulcis Peel Oil	10 Drops* (≈ 0.5 grams)	0.33%
	Totals	150.5 grams	100.00%

Information about ingredients to include in the PIF for Sweet Orange Lip Balm. The PIF details the common name, INCI name, the quantity of each ingredient used by percentage in the final product.

Cosmetic Product Safety Report

A key component in the PIF is the Cosmetic Product Safety Report (CPSR) issued by a qualified safety assessor. To determine if a cosmetic is safe, an assessor such as myself will calculate the Margin of Safety (MoS) for each ingredient in the formulation. Each ingredient in Sweet Orange Lip Balm has a sufficient MoS to declare the product safe.

Submit a Cosmetic Product Portal

With the RP and PIF in place, the cosmetic product is then listed on the new UK Submit a Cosmetic Product Portal. A description, detailed formulation and pictures of the product with labels are listed on the portal. Once submitted, your product is assigned a number, and you can legally sell your cosmetic in the UK.

Labelling Sweet Orange Lip Balm

Correct labelling is a must if you would like to sell your beeswax lip balm. The required information can be printed on one label or more, as long as the correct information appears on the product packaging. Figure 10 displays those prepared for Sweet Orange Lip Balm; an eye-catching top label and a more informative bottom label. On the left, the name of the product and on the right, the cosmetic's ingredients.

Further obligatory information is on the right, including the INCI name of ingredients, listed in descending quantity (see Table 3). Brassica Campestris Seed Oil (rapeseed oil) is the largest percentage at 59.8% of the ingredients in Sweet Orange Lip Balm, so it is listed first. The remaining ingredients are in descending order, followed by the allergens. Citrus Aurantium Dulcis Peel Oil makes up 0.34% of the final product. The allergens are constituents of the essential oil and are listed on the label if they meet the threshold for declaration.

The threshold for allergens is 0.01% for rinse-off products and 0.001% for leave-on cosmetics. Sweet Orange Lip Balm is applied and not rinsed off - a

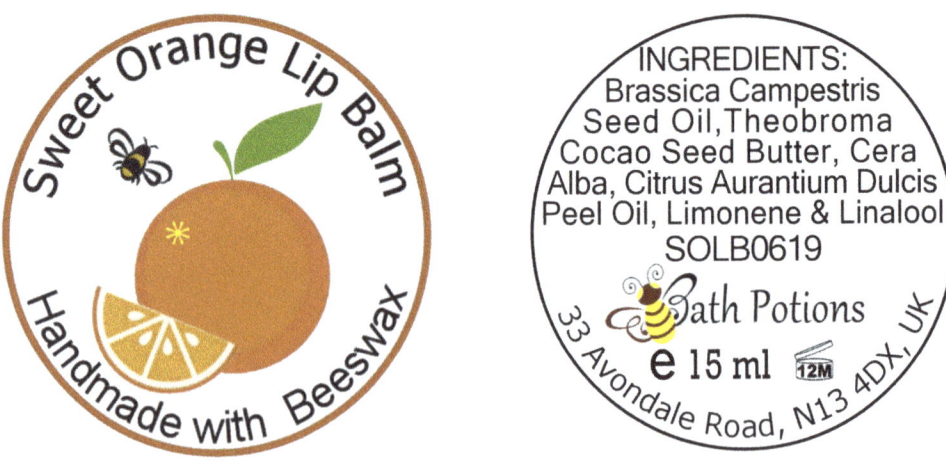

Figure 10. *Sweet Orange Lip Balm labels designed for the top and bottom of a 15 ml aluminium tin. The legally required information must be included on the labels.*

leave-on cosmetic. Allergens in the lip balm at a percentage equal to or above 0.001% must be on the label. If allergens are present below 0.001% in lip balm, they do not need to be specified on the label. The absence of allergens on a label does not mean the product is free from allergens; merely the allergens are present below the threshold for reporting.

Sweet Orange Lip Balm contains three allergens (Table 4), all of which are constituents of the sweet orange essential oil. Cosmetic ingredients, including allergens, are assigned a unique numerical identifier by the Chemical Abstracts Service (CAS) to chemical substances. The supplier's allergen declaration sheet indicates citral is present at 0.15%, limonene 95% and linalool 0.40%. The allergens in Sweet Orange Lip Balm above the threshold for reporting are limonene at 0.3156% and linalool at 0.0013%. You will note both of these allergens appear on the label after the primary ingredients.

Table 4. *Total Allergen percentage in Sweet Orange Lip Balm*

Allergen	CAS Number	% in Essential Oil	% in the Final Product
Citral	5392-40-5	0.15%	0.0005%
Limolene	5989-27-5	95.00%	0.3156%
Linalool	78-70-6	0.40%	0.0013%
		Total	0.3174%

The recipe above uses 0.5g of sweet orange essential oil. The percentage of each allergen in the final product is above. Limonene and linalool are present at greater than 0.001% and therefore are required to be listed on the label, while citral is not.

Below the ingredients is the batch number SOLB0619. The batch number should be unique for each batched produced and noted in your records. For the lip balm recipe in this booklet, the batch number begins SOLB to indicate Sweet Orange Lip Balm followed by 0619 to indicate June of 2019. You do not need to use a date; you could use a sequence of numbers. For example, SOLB001 for the first batch and SOLB002 for the second and so on.

The Responsible Person (in this case, a legal entity, Bath Potions) must list their name along with their address. Further information required is the nominal content of 15 ml for the packaged Sweet Orange Lip Balm. The label must also indicate the date minimum durability or a period after opening symbol. This lip balm should be used within 12 months of opening, as noted by the open jar symbol stating 12M.

> **Formulator's Tip**
> Most of us recognize the spelling cocoa butter. Some of the INCI names for products from the cocoa plant are spelled differently. Note the INCI spelling of Theobroma Cacao Seed Butter

Selling Your Lip Balm

After completing all of the steps above, your Sweet Orange Lip Balm is ready to sell. The finished product is a moisture-rich and protective lip balm with a subtle chocolate orange flavour (Figure 12).

Sweet Orange Lip Balm makes a fantastic addition to a market stall selling honey and other hive products. While the requirements to comply with legislation may seem intimidating, this booklet provides the recipe and outlines what you need to complete to sell in the UK. As a safety assessor, I can write your Cosmetic Product Safety Report (CPSR) for Sweet Orange Lip Balm and help you complete the other steps. Why not make and sell your own Sweet Orange Lip Balm?

Figure 11*. Sweet Orange Lip Balm in aluminium pots. Made with a blend of rapeseed oil, cocoa butter and beeswax, labelled and ready to sell.*

References

Robb, SJ. Dr Sara's Honey Potions. Northern Bee Books, 2009.

Robb, SJ. Beauty and the Bees. Northern Bee Books, 2012.

Robb, SJ. Making and Selling Cosmetics: Beeswax Lip Balm. BBKA News; September 2019, p298-301.

Robb, SJ. Making and Selling Cosmetics Honeycomb Cleansing Cream. Northern Bee Books, 2020.

Official Journal of the European L 342, 22.12.2009, Regulation (EC) No 1223/2009 of the European Parliament and of the Council of 30 November 2009 on cosmetic products

Regulation (EC) No 1223/2009 Annex I, Cosmetic Product Safety Report

Regulation (EC) No 1223/2009 Annex II, List of substances prohibited in cosmetic products

Regulation (EC) No 1223/2009 Annex III, List of substances which cosmetic products must not contain except subject to the restrictions laid down

Regulation (EC) No 1223/2009 Annex IV, List of colorants allowed in cosmetic products

Regulation (EC) No 1223/2009 Annex V, List of preservatives allowed in cosmetic products

Regulation (EC) No 1223/2009 Annex VI, List of UV filters allowed in cosmetic products

Index

A
Allergens 19, 20
Almond Oil 6, 9
Aroma 11

B
Balm 7, 9, 11
Batch Number 21
Beeswax (see Cera Alba) 6, 9, 10, 11, 13, 15, 18, 19, 22
Body Temperature 10
Brassica Campestris Seed Oil (see Rapeseed Oil) 18, 19, 20
Brexit 16
Butters 9, 10

C
Cera Alba (see Beeswax) 18, 20
Chemical Dossier 17
Chemistry of Mixtures 8
Citrus Aurantium Dulcis Peel Oil 18, 19, 20
Cleansing Cream 7
Cocoa Butter 6, 9, 10, 11, 13, 14, 15, 18, 21, 22
Cosmetic Allergens 23
Cosmetic Product Safety Report (CPSR) 17, 19, 21

D
Drops 13, 15, 18

E
Emollient 6, 11
Epidermis 11
Equipment 7, 11, 12, 13
European Union 4, 7, 16

F
Film-forming 11
Flavour 9, 10, 15, 21
Formulator's Tips 10, 13, 18, 21
Functions of Ingredients 11

H
Honey 7, 21

I
Ingredients in Sweet Orange Lip Balm.. 11, 19
International Fragrance Association (IFRA).. 18
International Nomenclature of Cosmetic Ingredients (INCI).............. 18, 19, 21

L
Labelling Sweet Orange Lip Balm.. 19
Legal Requirements 16
Lipophilic...9, 10

M
Making the Sweet Orange Lip Balm 13
Margin of Safety... 19
Melting Temperature.. 10
Method of Manufacture..8
Mixtures 8, 9, 10
Mineral Oil..9

N
Nominal Content.. 21
Non-aqueous 18

O
Oils 9, 10

P
Packaging.. 12, 13, 19
Paraffin Wax...9
Percentage... 18, 19, 20, 21
Period After Opening 21
Product Information File (PIF)... 16, 17, 18, 19

R
Rapeseed Oil...9, 10, 11, 13, 14, 15, 18, 19, 22
Recipe for Sweet Orange Lip Balm 13
Responsible Person (RP).. 16, 17, 19, 21

S
Scale 11, 12, 13, 14
Selling Your Lip Balm 21
Skin Conditioning 11
Submit a Cosmetic Product Portal 17, 19
Sweet Orange Essential Oil (see Citrus Aurantium Dulcis Peel Oil) ... 6, 10, 11, 13. 15, 18, 20, 21

T
Theobroma Cocao Seed Butter 18, 21
Threshold for Declaration of Allergens 19, 20

W
Water 10
Waxes 9, 10

Lightning Source UK Ltd.
Milton Keynes UK
UKHW050003091021
391767UK00004BA/48